# ENGINEERING MARVELS OF AUSTRALIA

# Australia's Hydro-Electric Power Schemes



Alison Hideki

Redback Publishing
PO Box 357
Frenchs Forest NSW 2086
Australia

www.redbackpublishing.com.au
orders@redbackpublishing.com.au



978-1-922322-10-4



Author: Alison Hideki
Editor: Michael Anderson
Proofing: Marianne Lindsell
Designer: Redback Publishing

Original illustrations © Redback Publishing 2020
Originated by Redback Publishing

Printed and bound in China

Acknowledgements
Abbreviations: l—left, r—right, b—bottom, t—top, c—centre, m—middle
We would like to thank the following for permission to reproduce photographs: With thanks to National Library of Australia and Thiess Bros p8, 13, 14, 15, 16 (Images © shutterstock) p6b - Snowy Mountains by Pee Tern via Wikimedia commons, p8b - National Library Australia, Work carried out on Snowy Mountains project by Thiess Bros, 1961-1968, PIC/13578/159-699 LOC Album 1143, p8b - Geehi Dam area rock processing plant, Snowy Mountains, 1962, PIC/13578/355 LOC Album 1143, p9b - Tellurometer via http://www.militarysurvey.org.uk, p9b - Workers placing reinforcement in tunnel outlet-inlet structure, 1964, PIC/13578/626 LOC Album 1143, p12 - Eucumbene Dam by Picturesk via Wikimedia commons, p13m - Caterpillar 650 scraper passing through Bella Vista, 1962, PIC/13578/593 LOC Album 1143, p13b - Three 30-ton Euclid trucks passing through Bella Vista, 1962, PIC/13578/353 LOC Album 1143, p13t - Workers at the completion of excavation in Snowy-Geehi Tunnel, 1964, PIC/13578/650 LOC Album 1143, p14b - Workmen removing drills prior to loading at Snowy adit near Geehi Dam, 1962, PIC/13578/178 LOC Album 1143, p14t - Excavating diversion tunnel outlet portal, 1962, PIC/13578/252 LOC Album 1143, p15b - Sliding floor and drilling jumbo on mobile platform in Snowy-Geehi Tunnel looking upstream, 1962, PIC/13578/312 LOC Album 1143, p16b - Sliding floor and jumbo mobile drilling platform in Snowy-Geehi Tunnel as seen from tunnel face, 1962, PIC/13578/311 LOC Album 1143, p17t - Guthega power station and Inside the machine hall by SkildCaddy via Wikimedia commons, p21b Pipe line & power station, Waddamana, ca 1900, p22b - Lake Rowallan by www.4wdgirl.wordpress.com, p30b - Cabramurra by Picturesk via Wikimedia commons



NATIONAL LIBRARY OF AUSTRALIA
A catalogue record for this book is available from the National Library of Australia

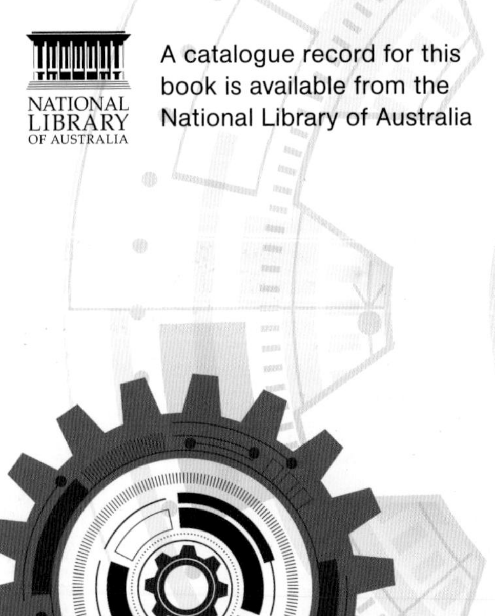

# Contents

# What is Hydro-Electric Power?

Hydro-electric power is electricity produced from the energy of moving water. In a hydro-electric power station, the energy of moving water is converted to electrical energy.

## HOW A HYDRO-ELECTRIC POWER STATION WORKS

In most cases, a dam is built across a river to stop the flow of the river and hold a large body of water behind it. The water is released through large pipes from the dam and it flows downhill to the power station. The steeper the pipes, the faster the water flow. When the water reaches the power station, it flows over the blades of turbines, spinning the turbines at great speed. The spinning turbines are connected to generators that each contain a very strong magnet. The spinning turbines cause coils of wire cable in the generator to spin around the magnet. This produces electricity in the wires. The electricity then travels through power cables to be used in hundreds of different ways in homes and in industry.

### What is an Engineering Marvel?

An engineering marvel is a project that breaks new ground and solves problems that might have seemed too difficult to overcome. Often, new building methods are employed in the construction of an engineering marvel. The invention of new machines and new technology often leads to the creation of new engineering marvels.

## AUSTRALIAN MARVELS

Australia's hydro-electric projects are amongst the most astonishing feats of engineering to be found anywhere in the world. The Snowy Mountains Hydro-electric Scheme, spanning Victoria and New South Wales, remains one of the great engineering feats of modern times. Built between 1949 and 1974, this mammoth project is still the largest engineering project ever undertaken in Australia. Other hydro-electric projects such as the Gordon Dam and the Reece Dam in Tasmania and the Ord River Scheme in Western Australia are also examples of Australia's engineering marvels.

### Technology Time Machine! Waterwheels

People have used the power of moving water for thousands of years. One of the most common ways in which waterpower is harnessed is by use of the waterwheel of a mill. A huge wheel with flat spokes is built over a stream of river. As the water passes over the flat spokes, the wheel is turned. An axle from the wheel is connected to a huge millstone inside the mill beside the stream. The axle turns the stone and the stone grinds the grains of wheat in a trench below it to make flour. The mill wheel is a type of turbine. The first hydro-electric power stations were built on the model of the waterwheel. Mills with waterwheels were used in China as far back as 3,000 years ago.

## PROBLEMS

In order to provide a large body of water for hydro-electric power stations, it is usually necessary to construct a dam, or dams. Dams flood the landscape, destroying the habitats of many animals and changing the natural environment forever. A second problem for hydro-electric power stations is their dependence on rainfall. In times of drought, the water source for the hydro-electric power station can be reduced, affecting the amount of electricity that can be produced.

## HOW MUCH ELECTRICITY?

The amount of electricity a power station produces depends on the number of turbines and generators it contains. The power station operators can regulate the amount of electricity being produced by closing down or by opening more turbines.

# The Snowy Mountains Hydro-Electric Scheme

The Snowy Mountains Scheme is recognised as one of the seven civil engineering wonders of the modern world. It grew out of an idea for providing irrigation for crops in dry areas of Victoria and New South Wales.

## THE SNOWY MOUNTAINS

In the Snowy Mountains of south-eastern New South Wales and north-eastern Victoria, the Snowy River, as well as other rivers and creeks, carry the melt-water of the snow that falls in the winter months down the valleys to the plains below. For many years, engineers spoke of the possibility of capturing water from water-rich regions, such as the Snowy Mountains, and channelling that water to drier regions for irrigating land that could then be used to grow crops.

## THE PLAN

In 1946, a plan to divert the waters of the Snowy River into the Murrumbidgee and Murray Rivers was drawn up in Canberra. The Federal Minister for Works and Housing at that time, Nelson Lemmon, thought the plan was brilliant, but he wanted to include a large system of hydro-electric power stations. The power stations would provide enough electricity to satisfy the demands of householders and of industry for 50 years to come. The water used in the power stations would flow on into the Murray and Murrumbidgee Rivers, allowing millions of hectares of dry land to become productive through irrigation. At that time, there was little consideration of how the scheme would affect the Snowy River.

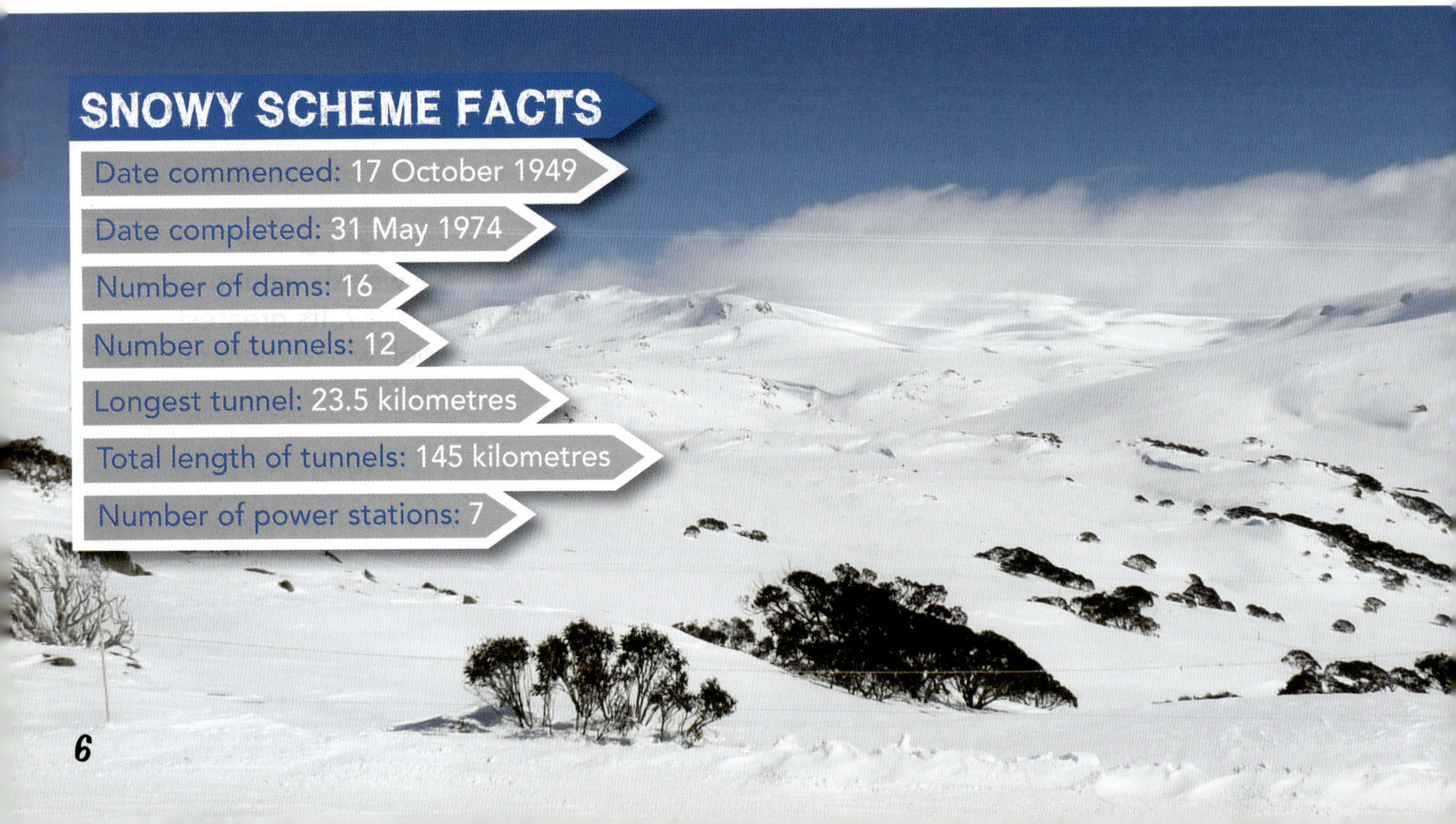

## SNOWY SCHEME FACTS

- Date commenced: 17 October 1949
- Date completed: 31 May 1974
- Number of dams: 16
- Number of tunnels: 12
- Longest tunnel: 23.5 kilometres
- Total length of tunnels: 145 kilometres
- Number of power stations: 7

## Technology Time Machine! The Slide Rule

When the Snowy Scheme was being planned and built, there were no computers to work out complex equations. Snowy engineers relied on a slide rule more than any other device.

## CONVINCING THE PEOPLE

Many experts doubted that the scheme would ever be built. It was thought that rivalry between New South Wales and Victoria would kill the Scheme before it got off the ground. But Australia had just emerged from World War II - a war in which the invasion of Australia by Japan had been a real threat. Nelson believed that he could convince the states and the Australian people that the Snowy Mountains Hydro-electric Scheme was a matter of national security – and that it would help to make Australia a strong nation, capable of defending itself against any future attack. Lemmon's instinct proved accurate and in 1949 work began on the Snowy Scheme.

## AUSTRALIA IN 1949

Australia was not a wealthy country by world standards in 1949. Its greatest resources were wool, wheat and iron ore. Nations such as Britain and the USA had many more industries than Australia. After the shock of a near invasion by the Japanese in World War II, Australians came to realise the importance of self-reliance, of building up industries and making Australia a more powerful nation. The Snowy Scheme, with its promise of unlimited electrical power for industry, made sense to a nervous nation.

# A Design Challenge

The team of engineers that designed the dams, bridges, tunnels and power stations of the Snowy Mountains Hydro-electric Scheme had to overcome problems that had never before been encountered in an engineering project in Australia. Many of the tunnels were to be longer and deeper than any bored in Australia up to 1949, and they had to be bored safely. Bridges had to be built in wild country over raging torrents. But the biggest problem that the engineers faced was that of precision.

## PRECISION ENGINEERING

The Snowy project called for the construction of 12 huge tunnels, 16 dams, 30 aqueducts and seven power stations spread over 100,000 square kilometres. Just the construction would be difficult enough, but the project was to be built in rugged mountain terrain covered in forest, and it had to all come together as precisely as the pieces of a jigsaw puzzle. Tunnels would begin on opposite sides of a mountain and would have to meet up in the middle. This called for very accurate calculations, as the two teams tunnelling could not be sure of the exact position of the other. The volume of water each of the dams would hold had to be worked out accurately.

## SURVEYING THE LAND

Before any work could begin on the Snowy project, thousands of square kilometres of countryside had to be surveyed so that the engineers could know the exact height of mountains, hills and ridges, and the exact volume of water flowing along hundreds of creeks, streams and rivers. It was also necessary to know about the soil and rocks of the region.

Geehi Dam area rock processing plant

## INSTRUMENTS FOR MEASURING

In order to make precise measurements of mountains and hills, special instruments had to be used. Some of these instruments were developed especially to help surveyors make precise measurements on the Snowy project. Surveyors relied on instruments such as theodolites, tellurometers and aneroid barometers to measure height and distance. New and more accurate versions of these instruments helped surveyors make the pinpoint measurements that the Snowy project required.

### Technology Time Machine! The Tellurometer

TL Wadley, a South African engineer, developed a device for use on the Snowy project that helped save valuable time in surveying. The tellurometer measured distances by bouncing electromagnetic waves off a reflector, then measuring the time that the reflected waves took to return to the device.

## ROCKS PLUS BOLTS

Safety was a very important consideration in the building of the Snowy Scheme. Tunnelling was especially dangerous for workers and tunnels often collapsed. On the Snowy project, a method of rock bolting was developed to ensure that huge rocks did not break free from the tunnel wall. Deep holes were bored through dangerous rocks in the tunnel walls and long bolts, metres in length, were driven into the holes to hold rocks in place.

# The Dams

The Snowy Mountains Scheme involved the construction of 16 dams. Some of the dams, such as the Jindabyne Dam and Eucumbene Dam, were immense. The Eucumbene Dam is 116 metres high and 580 metres long. Building the main wall of a large dam is a very difficult engineering task. Designers have to make sure that the wall will be strong enough to withstand the great pressure of the water behind it. Building a dam wall is like building a mountain, but a mountain that is stronger than a real one.

## TYPES OF DAMS

Dams are built in different ways according to the land on which they stand and the volume of water they must hold back. On the Snowy Mountains Scheme, four types of dam wall construction were used:

The embankment dam wall was used when a very long wall was required and when a great volume of water was to be held back. The Talbindo Dam, the largest dam of the Snowy Scheme, is an embankment dam.

### CONCRETE ARCH DAM

A dam made of a wall of concrete that curves into the water behind it

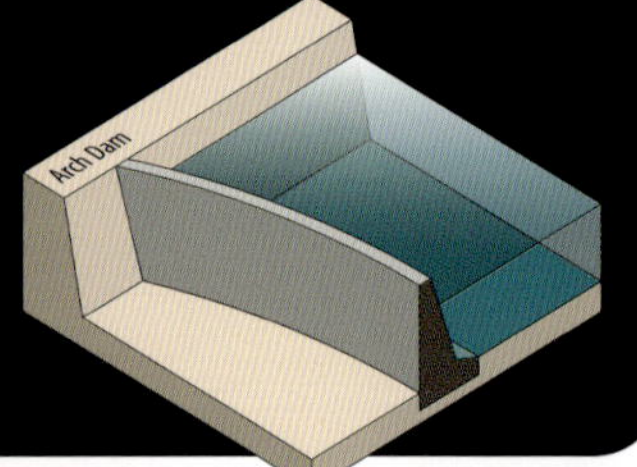


### EMBANKMENT DAM

A dam filled with earth or rock - also called a rockfill or earthfill dam

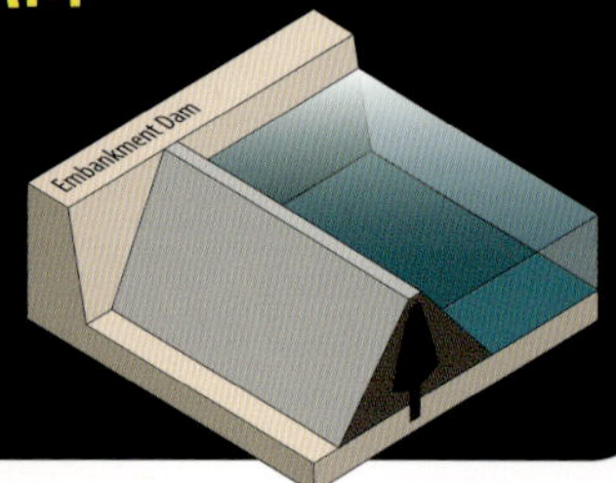


### CONCRETE GRAVITY DAM

A type of dam in which the concrete wall is curved back towards the weight of the stored water

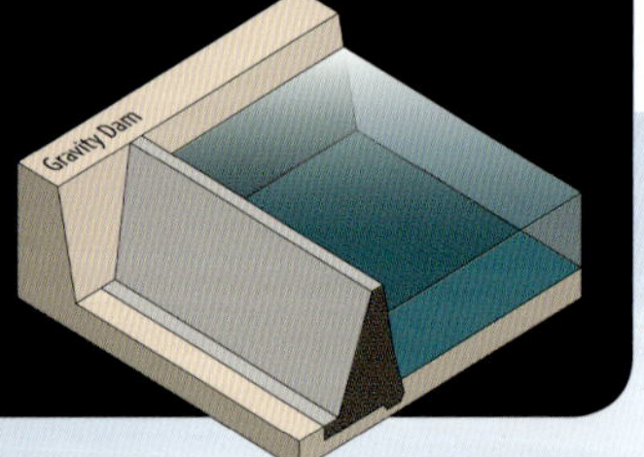


### CONCRETE BUTTRESS DAM

A concrete dam wall supported by wedge structures built against it (called buttresses)

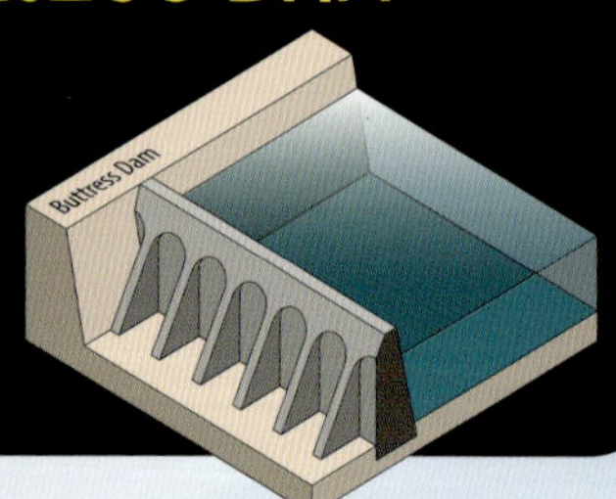

## The Right Dam in the Right Place

Earthfill and rockfill dams were built whenever a very long and high dam wall was required. Concrete gravity dams, concrete arch dams and slab and buttress dams were used when the natural terrain provided a narrower outlet for a reservoir – an opening between two hills, for example. Sometimes concrete dam walls were built as spillways, with gates in them that opened to release water from the reservoir. Earthfill and rockfill dams usually included a concrete spillway.

## EUROPEAN IMMIGRANTS

In 1948, when the very first preparations began for the building of the Snowy Scheme, Australia did not have a large enough workforce to complete the project. The Federal Government assisted both skilled and unskilled workers from Europe to migrate to Australia to work on the Snowy Scheme. Many migrants brought their families with them. Towns on the fringes of the project, such as Cooma in New South Wales, became home to migrants and their families. The migrants came from Italy, Germany, Belgium, The Netherlands, Britain, Latvia, Lithuania, Estonia, Ukraine, Russia, France, Switzerland, Poland and Hungary.

# The Eucumbene Dam

## BUILDING THE EUCUMBENE DAM

The Eucumbene Dam took only two years to construct (1956-1958), but preparations began much earlier, in October 1949. First, the water of the Eucumbene River had to be diverted by a tunnel. Roads were bulldozed to the site of the dam wall, and then blasting with high-explosives opened a deep trench where the dam was to be built. Massive trucks called antars, carrying more than 100 tonnes of rocks and earth in each load, moved slowly up and down the steep mountain roads. The trench, which was almost 600 metres long and 686 metres wide, was packed with crushed stone, which became the foundation of the embankment dam wall.

Clay carried to the site from a quarry was packed onto the foundation to make the base of the dam wall. Tens of thousands of truckloads of earth and stone were packed onto the clay and levelled out by huge bulldozers. The wall grew higher and higher in terraces, one layer of the wall sitting on top of a wider layer, up and up for 116 metres.

Concrete was not used to construct very long and high dam walls, such as the wall of the Eucumbene Dam, as concrete is a much more expensive building material than clay and rock.

## SKILLED WORKERS

Approximately 170,000 people worked on the Snowy Scheme over the 25 years it took to complete. Not all of those people were at work on the project at the one time, but over the first ten years of construction the workforce grew. Some of the workers were specialists, such as engineers, mechanics, welders and explosive experts, while others were labourers.

## SAFETY

Many of the building practices employed on the Snowy Scheme would be considered too dangerous today. Trucks drove on bulldozed roads that sometimes gave way in the rain. Explosives were sometimes ignited without sounding a warning siren. The huge trucks and machinery did not have backing beepers, as they have today. Men worked without the safety gear they wear today, such as eye and ear protection, and facemasks to filter the dust and grit before it entered their lungs.

121 men died as a result of accidents on the Snowy Scheme. Thousands more were injured, often quite badly.

## THE BIG MACHINES

Building the Snowy Scheme was mostly about moving millions of tonnes of earth and rock. The biggest heavy-moving machinery and vehicles ever seen in Australia were used to accomplish these tasks. Trucks such as the massive antars came from Britain; bulldozers came from the USA. In the middle and later years of the project, some machinery was imported from Germany and Italy.

# Boring the Tunnels

Altogether, 145 kilometres of tunnels were bored on the Snowy Scheme, often through solid rock. The longest tunnel, running between Eucumbene and the Snowy River, was 23.5 kilometres long. A total of 12 tunnels were built between 1955 and 1969.

The tunnels were huge. The largest of the tunnels, such as the Eucumbene-Tumut Pondage Tunnel, was 6.30 metres in diameter - as high as a house and big enough for trucks to drive through.

## BLASTING THROUGH ROCK

Tunnelling through rock requires explosives. Sections of the tunnels were blasted open, and then the rock debris was cleared with machinery and by hand. In some of the Snowy tunnels, gigantic boring machines were used to drill the opening. Work was carried out in shifts that sometimes ran 24 hours a day. Hundreds of men could be at work in the largest tunnels at one time. Some of the workers had the job of rounding the walls of the tunnel, using jackhammers or sometimes using traditional mining tools such as picks. Other workers cleared rock debris after blasting.

*Rock drill blasting through rock*

Loaders filled the trucks and trolleys that carried millions of tonnes of rock debris out of the tunnels. Boulders in the walls of the tunnel were secured in place with steel bolts metres in length. Once the tunnels were bored, sections of the walls were sealed with concrete and steel plate.

## JOINING UP IN THE MIDDLE

Most of the tunnels of the Snowy Scheme were bored through mountains. In order to hasten the work, tunnelling teams would begin work on opposite sides of a mountain and join up in the middle. Measurements were taken a number of times each day to ensure that the tunnel halves were running at the right angle, at the right depth and in precisely the right direction. Of the 12 Snowy tunnels, only one went off-course, and that one was corrected as soon as the mistake was seen.

## Technology Time Machine! The Sliding Floor

A world-first on the Snowy Scheme was the use of the sliding tunnel floor. This was a steel platform of three levels mounted on rails that carried workers, drilling equipment and other machinery along the route of the tunnel to the point where the day's work was to commence. Workers could carry out operations on the whole of the tunnel wall from the platform. The three levels allowed a greater number of men to work in a tunnel at one time. The sliding tunnel was, in fact, a moving scaffold. Valuable time was saved because traditional scaffolding did not have to be erected and taken down at every new stage of digging.

## HOW LONG DID IT TAKE?

The tunnels of the Snowy Scheme were bored in world record time. The 165 metres of the Snowy-Geehi tunnel were bored in one week in 1963, a record for solid rock tunnelling. The longest tunnels took up to five years to complete, while others took months. The speed of tunnelling depended on the rock that had to be penetrated, and the dangers involved.

*Sliding tunnel floor*

# Building the Power Stations

The Snowy Scheme consists of seven power stations. Two were built underground.

## EASY AND HARD

Building the power stations was in many ways an easier task than building the dams and tunnels of the scheme. A hydro-electric power station is simply a sturdy building housing large items of machinery, turbines and electrical systems with gates that allow water from the tunnels to enter and exit.

Building the stations in the rugged country of the Snowy region made the task difficult. The base of a power station has to be much lower than ground level, so blasting was required to excavate the deep holes in which the foundations of the stations would be laid. Once part of the walls and floors of the power stations had been built, the giant turbines and other large items of machinery had to be installed. Some of the turbines and transformers weighed more than 100 tonnes. Very large trucks were required to carry the machinery to the building sites, with two trucks pulling the load and one truck pushing. Unloading the machinery from the truck bed was particularly difficult. Block and tackles were used to lift the loads from the trucks. Cranes and hydraulic jacks were used to move the turbines and transformers into position.

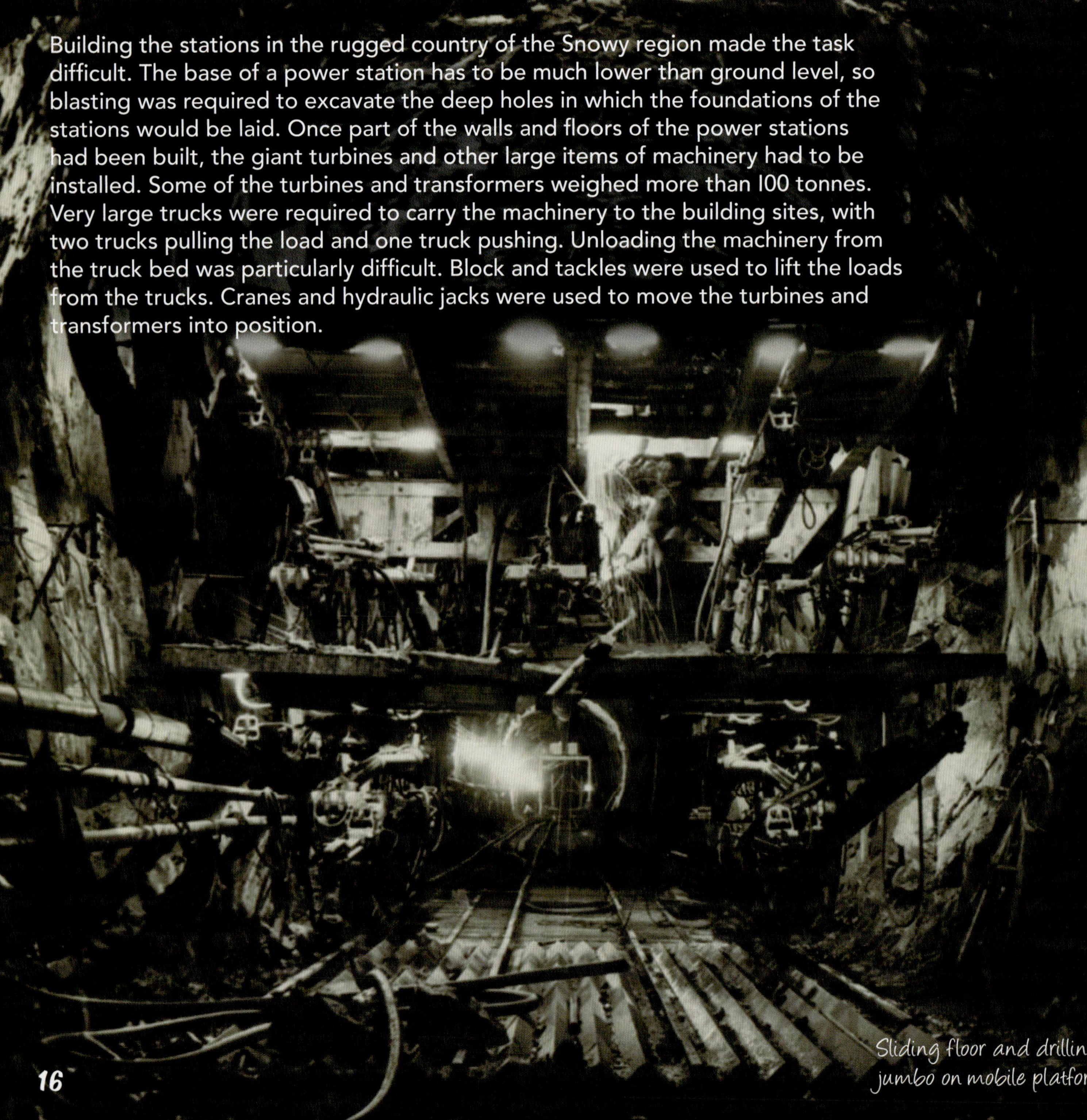

*Sliding floor and drilling jumbo on mobile platfor*

Guthega Station

## SWITCHYARDS

While the power stations were being built and generating equipment installed inside, work was also being carried out on the switchyards outside the stations. Switchyards collect the electrical power generated by the turbines and feed it into power cables. The electricity then travels through the power cables to cities and towns.

### Technology Time Machine! Block and Tackle

Block and tackle technology was used for heavy lifting. Block and tackle and pulley technology is hundreds of years old. These days, special cranes are used. Although cranes were used on the Snowy Scheme, they were fixed in place. Modern cranes are mobile and can move from site to site.

## UP AND RUNNING

The first of the Snowy Scheme power stations, the Guthega Station, began operation in 1955. It had taken five years to build, including preparations such as building roads and the clearing of the building site. The six power stations that followed Guthega were built more quickly as the workforce grew and more roads were bulldozed. The second power station, Tumut I, began operation in 1959. It wasn't until 1974 that all power stations were in full operation.

## THE SNOWY

The Snowy Mountains Hydro-electric Power Scheme is Australia's largest hydro-electricity generation scheme and largest provider of water for irrigation. It produces 32 per cent of all renewable energy in the mainland National Electricity Market. It provides electricity to New South Wales, Victoria, Queensland, ACT, and South Australia.

In 2017, Prime Minister Malcolm Turnbull announced plans to expand the Snowy Scheme. The expansion has been dubbed Snowy Hydro 2.0, and will increase the amount of electricity by 50 per cent.

## SNOWY HYDRO

The Australian Federal Government and the State Governments of New South Wales and Victoria own the Snowy Mountains Hydro-electric Power Scheme. Snowy Hydro is the name of the company that operates the Snowy Mountains Hydro-electric Power Scheme today. It is responsible for maintaining the dams, tunnels and power stations, as well as the delivery of electricity. About 650 people work for Snowy Hydro.

## THE ENVIRONMENT

Today people are much more concerned about the environment than they were at the time the Snowy Scheme was built. The Snowy Scheme has had a huge impact on the habitats of native animals. Snowy Hydro now works with the National Parks and Wildlife Service (NPWS) to protect the remaining areas of animal habitats.

With the announcement of an expansion of the scheme, conservation groups are questioning the project's potential impact on the area, including introduced animals and wetland loss.

## THE SNOWY RIVER

Before the building of the scheme, the Snowy River often ran at maximum height during the spring thaw, when melting snow filled the river. As demand for water for irrigation increased between 1975 and 2000, more and more water was diverted from the Snowy into the dams, and on into the Murray and Murrumbidgee Rivers, gradually reducing the downstream flow of the Snowy to one per cent of its capacity. The Snowy became an almost dry riverbed. A campaign to restore the flow resulted in more water being returned to the Snowy. By early 2017, the river's flow met an 18-year target of being reinstated to 21 per cent of its natural state.

## MAINTENANCE

Engineering works require constant attention, and are repaired all the time. The tunnels are checked regularly for faults. The dams are tested with instruments that can record any movement that may have taken place in the walls or if any decay has occurred deep inside the walls. Electrical engineers are constantly servicing the power stations. Turbines, transformers and generators are repaired or replaced when they become worn. Concrete can also rot or decay so engineers watch out for that using ultrasound devices.

## National Heritage

The Snowy Mountains Hydro-electric Scheme was awarded National Heritage listing in 2001, recognising it as 'an extraordinary engineering feat'. National Heritage listings are given to both natural and constructed sites that are considered of such importance that they should be preserved for future generations of Australians to admire and enjoy.

# Tasmania's Hydro-Electric Power

Tasmania has 30 hydro-electric power stations. In 2006, Tasmania began exporting energy to the mainland. These power stations now provide 5 per cent of Australia's energy, and about 40 per cent of its renewable energy.

## THE ISLAND STATE

As an island state, Tasmania was for a long time at a disadvantage in industry compared with the mainland states. From the time that it was established as a British colony in 1825 until the beginnings of air travel in the 1920s, everything that was sent to and from Tasmania went by ship. When trains and trucks were taking goods across borders on the mainland, Tasmanians still had to rely on cargo ships crossing Bass Strait. Industrial development lagged behind that of the mainland states. In the late 19th century, when power stations began supplying electricity, the coal that fuelled the stations was expensive for Tasmanians to import.

## RICH IN WATER

In the early 20th century, hydro-electricity was developed and Tasmania found itself very rich in the natural resource required to produce it - water. Tasmania has the highest state rainfall in Australia and the most reliable waterways.

## Zinc and Zinc Ore

Zinc is a valuable metal. It is one of the metals that make up brass and it is mixed with iron to make certain kinds of steel. Zinc ore is the rock that zinc is found in. The process of extracting zinc from zinc ore is called smelting, which requires a lot of electricity. There are large deposits of zinc ore in Tasmania. Hydro-electricity made smelting the zinc possible, and mining the ore worthwhile.

Zinc factory

## ELECTRICITY FOR ZINC

Tasmanian hydro-electrical engineering began in 1909, when businessman James Hynde Gillies invented a method of extracting zinc from ores. The method promised huge development in mining and zinc production in Tasmania. The stumbling block was the need for abundant and cheap electricity. At that time, there was no hydro-electricity in Tasmania. Over the following years, the Tasmanian Government set up a hydro-electric department to organise the building of hydro-electric power stations.

## QUEENSTOWN

There are large deposits of zinc ore at Queenstown in west Tasmania. In 1911, the Mount Lyall Mining and Railway Company selected Queenstown for a large scale ore-smelting operation. The smelting plant and the hydro-electric power station built to power it were completed in 1914 and are still in use today.

## THE FIRST HYDRO-ELECTRIC POWER STATIONS

The first power stations used natural reservoirs such as lakes as water storage. The Waddamana Power Station, the first station owned and operated by the Tasmanian Hydro-electric Department (later called the Hydro-electric Commission), was opened in 1916. The electrical output was small (seven megawatts in 1916 increasing to 49 megawatts by 1929), but it served the mining industry well. More smallscale hydro-electric power stations were built, as the benefit to industry became apparent.

The Waddamana Power Station

# Dam Construction

Almost every type of dam design has been used in Tasmanian hydro-electric schemes.

## EMBANKMENT DAMS

Embankment dams are the largest kind of dams. They include earthfill and rockfill dams. They are used for large power stations, when a large body of water needs to be held back. Embankment dams are long walls built across a wide opening. They include a layer of concrete laid on the side of the wall that faces the water to prevent the water seeping into the embankment. The wall is built up in layers, each layer narrower than the one below. Rock, gravel, earth and clay are used to build the wall.

**EMBANKMENT DAM**

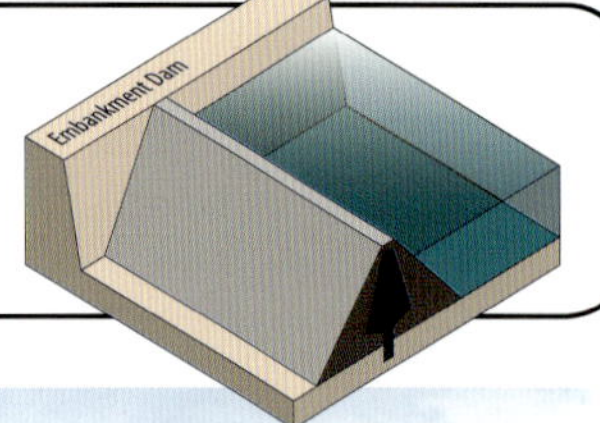


## CORED ROCKFILL DAMS

These are made up of a layer of clay sandwiched between two rockfill walls. The clay 'sandwich filling' is sometimes many metres thick. Clay is an excellent material for sealing water in a reservoir since it resists penetration. Like embankment dams, cored rockfill dams are used when a large wall is required, but they are cheaper to build than embankment dams.

### Cored Rockfill Dams Include:

- Rowallan Dam
- The Paragana
- The Echo
- The Wayatinah
- The Miena

Rowallan Dam

### The Reece Dam

The largest embankment dam in Tasmania is the Reece Dam, built between 1980 and 1984

## CONCRETE DAMS

There are three types of concrete dams, and all have been used in Tasmania. They are:

CONCRETE GRAVITY DAM

CONCRETE BUTTRESS DAM

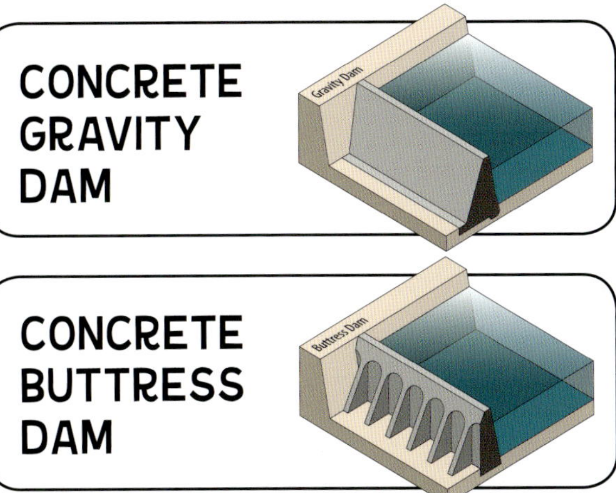

CONCRETE ARCH DAM

## CONCRETE GRAVITY DAMS

Concrete gravity dams use the weight of the concrete from which they are built to hold the water of the reservoir. Gravity dams are only used for small-scale projects. Tasmania's Henty, Liapootah, Pine Tier and Trevallyn are all concrete gravity dams.

## CONCRETE BUTTRESS DAMS

Like gravity dams, concrete buttress dams are used for small projects. A buttress dam uses a technique for making a structure stable developed a thousand years ago and used in churches and cathedrals. Wedge-like buttresses built against the dam wall support the wall of the dam. Tasmania's Meadowbank Dam is a concrete buttress dam.

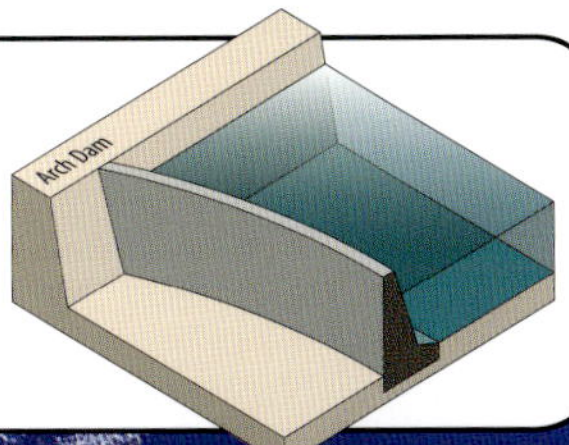

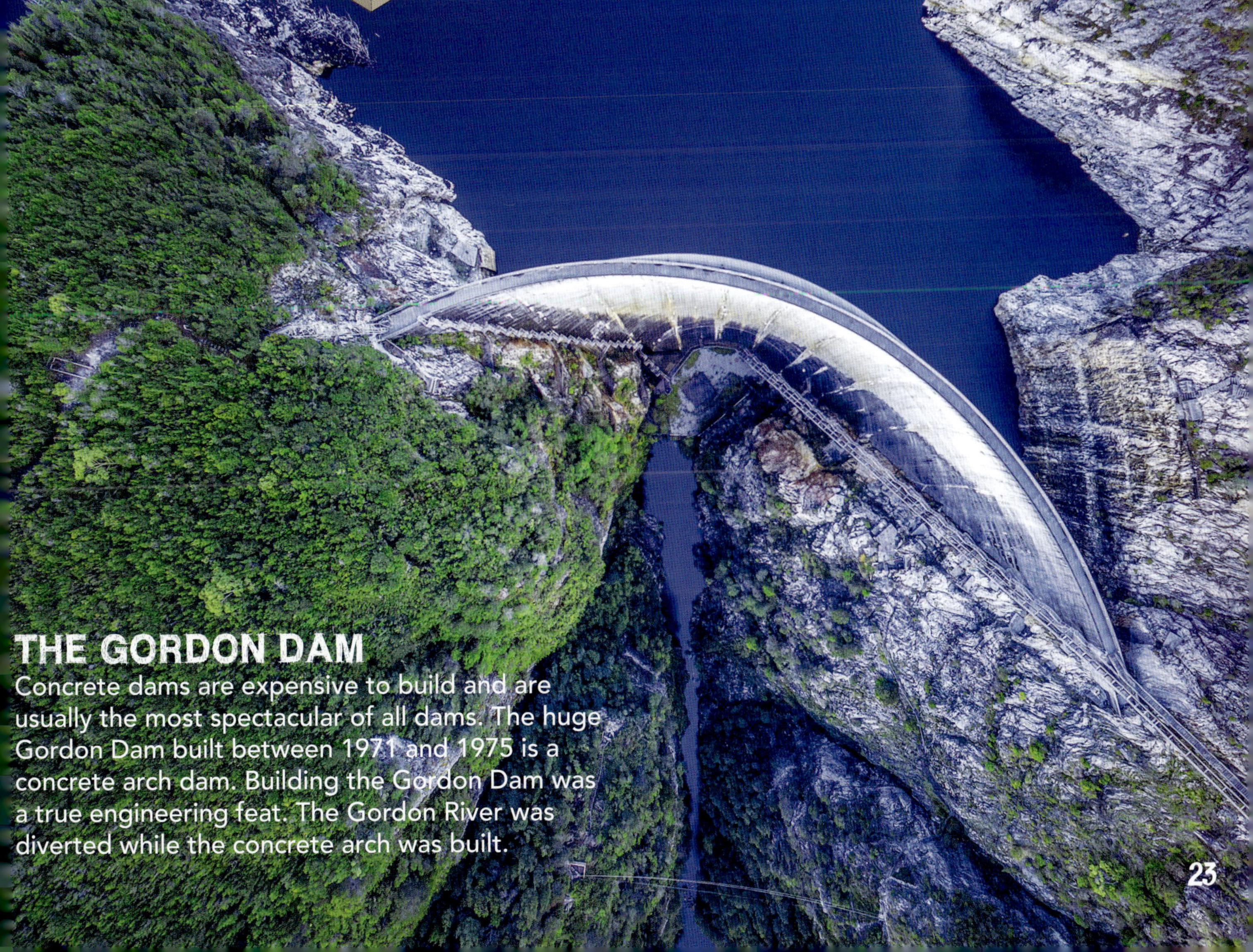

## THE GORDON DAM

Concrete dams are expensive to build and are usually the most spectacular of all dams. The huge Gordon Dam built between 1971 and 1975 is a concrete arch dam. Building the Gordon Dam was a true engineering feat. The Gordon River was diverted while the concrete arch was built.

# Dams in the Wilderness

Tasmania includes some of Australia's wildest regions, particularly in the west, the south west and on the central plateau. When the Tasmanian Hydro-electric Commission pushed ahead with plans for more and bigger dams and power stations between 1920 and 1938, it had to penetrate untamed wilderness in many cases, made up of mountains, dense forest and raging rivers.

## VILLAGES IN THE BUSH

Workforces of up to 1800 men had to be accommodated hundreds of kilometres from existing towns and cities. When the Commission first began these wilderness projects, no allowance at all was made for housing the workers' families. The Commission expected the workers to leave their families behind, or make personal arrangements. The Commission had difficulties in recruiting and keeping workers, so new plans were made that included building villages in the bush to house the workers and their families.

## Hydro Villages Today

Many of the villages built by the Tasmanian Hydro-electric Commission are still standing, dotted all over the map of Tasmania. These villages now have a smaller population than in past years, and house people working on the maintenance of completed schemes. In 2016, one whole village was up for sale.

## CONCERN FOR THE ENVIRONMENT

In the early years of Tasmanian hydro-electricity projects, large areas of untouched wilderness were submerged underwater as the dams held back the flow of rivers and huge lakes behind them. For many years, there was little concern for the environment and the Tasmanian Hydro-electric Commission had no branch responsible for assessing the environmental impact. In the past few decades, the now named Hydro Tasmania has developed a more detailed environmental program, although many people think more should be done to protect Tasmania's wilderness areas.

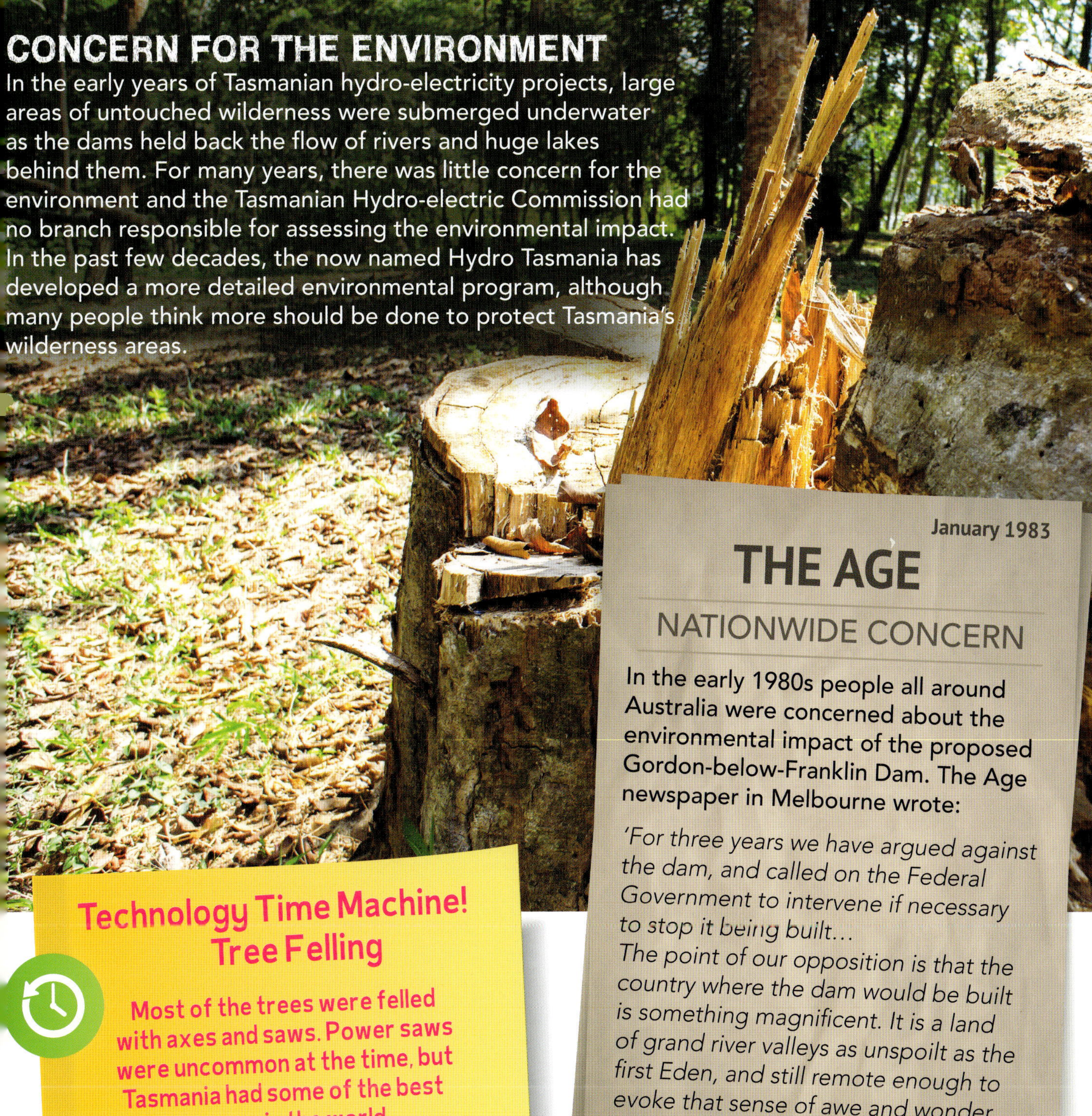

January 1983

### THE AGE

#### NATIONWIDE CONCERN

In the early 1980s people all around Australia were concerned about the environmental impact of the proposed Gordon-below-Franklin Dam. The Age newspaper in Melbourne wrote:

*'For three years we have argued against the dam, and called on the Federal Government to intervene if necessary to stop it being built…*
*The point of our opposition is that the country where the dam would be built is something magnificent. It is a land of grand river valleys as unspoilt as the first Eden, and still remote enough to evoke that sense of awe and wonder that the first explorers might have felt.'*

**Technology Time Machine! Tree Felling**

**Most of the trees were felled with axes and saws. Power saws were uncommon at the time, but Tasmania had some of the best axemen in the world.**

## ENGINEERS AND ENVIRONMENTALISTS

The massive Gordon Dam on Tasmania's Gordon River was the most controversial dam ever built in Australia. While the stunning concrete arch of the dam's retaining wall won international praise from engineers, the flooding caused by the damming of the river dismayed environmentalists. Tens of thousands of hectares of untouched native forest were submerged when the dam was filling. When the Tasmanian Government announced a second dam on the Gordon River – the Gordon-below-Franklin Dam – protests erupted all over Australia. The second Gordon dam would have submerged parts of the Franklin River and also many areas of pristine natural beauty. In 1983, a newly-elected Federal Labor Government prevented the dam from being built.

# Hydro-Electric Power Today

Tasmania's hydro-electric power stations produce 2618 megawatts of electricity, more than enough for Tasmania's electricity needs. Formerly known as the Hydro-electric Commission, or the Hydro, it is now called Hydro Tasmania and includes wind farms.

## BASSLINK

In April 2006, the 290-kilometre-long Basslink submarine cable running between Tasmania and Victoria was opened. The Basslink cable is a two-way transmission cable conveying electricity between Tasmania and Victoria. Since Tasmania is able to produce more hydro-electricity than it needs, and Victoria is seeking to boost its use of hydro-electricity, most of the transmission will be from Tasmania to Victoria. The Victorian electrical energy grid is linked to the energy grids of other mainland states, so Tasmanian hydro-electricity is being used across mainland Australia.

## MAINTENANCE AND NEW PROJECTS

Now that Tasmania is able to export electricity, more money is available for the upkeep and improvement of Tasmania's dams and power stations. Since November 2015, Hydro Tasmania has replaced large valves in the Trevallyn Dam.

Upkeep undertaken or planned:

- Tungatinah Power Station
- Tarraleah Power Station
- Paloona Power Station
- Meadowbank Power Station
- Fisher Power Station
- Cethana Power Station
- Rowallan Dam

*Water pipeline to Hydro-electric power station at Tarraleah*

# HISTORY AND HERITAGE

Hydro Tasmania closed some of the oldest power stations, such as those at Lake Margaret and at Waddamana. The oldest power stations and dams were built with technology that is now well out-of-date, such as wooden pipes. Some people living in these areas campaigned to have the power stations kept running. They claimed that the power stations and the dams that serve them are not only important economically to the community, but are also a vital part of Tasmanian history and heritage. The older dams have become tourist destinations for people wishing to see what hydro-electric engineering looked like before the age of high-tech.

Abandoned Hydro-electric Pump Station on Lake St Clair

Lake Margaret Hydro-electric power station

## UPKEEP

The maintenance of dams is mostly a matter of looking after the machinery used in opening and closing spillway gates and in generating electricity. The dam walls themselves do not require intensive maintenance. Regular checks are made to determine if the dam walls have moved, or if there is decay under the surface, but it's rare for a dam to require large-scale overhaul. However, the machinery, cables, transformers, turbines and generators of Tasmania's hydro-electric power stations are serviced all year round.

# Hydro-Electric Snapshot

## THE HUME, NEW SOUTH WALES

The Hume Reservoir in New South Wales was built between 1919 and 1931. It was one of Australia's biggest engineering projects before construction of the Snowy Scheme. The embankment wall of the Hume is 1.6 kilometres long and 51 metres high.

The Hume Reservoir was expanded in the 1950s. The entire town of Tallangatta was moved to a new location to allow for the rising waters. The Hume Dam is one of the few large-scale dams in Australia to have slipped on its foundations over the years. It was extensively repaired and renovated in the 1990s for fear of its collapse and the destruction of the Albury-Wodonga Township to the west.

## EILDON, VICTORIA

Lake Eildon on the Goulburn River in central Victoria was constructed as an irrigation storage dam and as a hydro-electric source between 1949 and 1954. Its water storage capacity is 3.5 million mega litres. The huge Eildon project was under construction at the same time as the Snowy Hydro-electric Power Scheme, and like the Snowy Scheme also employed thousands of European immigrants. The Eildon Dam is an embankment dam with a wall running for approximately 1.5 kilometres between Mount Sugarloaf and Mount Pinnegar.

# THE ORD RIVER SCHEME, WESTERN AUSTRALIA

The Ord River Scheme, in the eastern Kimberley region of northern Western Australia includes a hydro-electric generation component. The scheme was built between 1963 and 1972. Its construction created Australia's biggest artificial lake, Lake Argyle, which covers 741 square kilometres. The dam of Lake Argyle is an embankment dam built across the valley between two hills.

# WIVENHOE, QUEENSLAND

Lake Wivenhoe is a huge artificial lake created by an embankment dam. It is 80 kilometres from Brisbane and holds 2.5 times the capacity of Sydney Harbour, or approximately 1.5 million mega litres. The main purpose of creating the lake was to provide a reliable water supply for south-eastern Queensland, but Wivenhoe also includes a hydro-electric power station. To gain the height needed for power generation water is pumped from Lake Wivenhoe to a point above the power station before being released through a tunnel to the turbines of the generator.

# Hydro-Electric Tourism

There are many places you can visit to learn more about these incredible engineering marvels.

## SNOWY HYDRO

Visit the **Snowy Hydro Discovery Centre** in Cooma, NSW. The centre showcases all the amazing history from the early days of construction, up to today. See interactive and hands on displays that teach you about engineering, the environment, the National Electricity Market, history, water management and power generation.

The township of **Cabramurra** housed Snowy Scheme workers and was a base for Tumut 1 and Tumut 2 Power Station projects during the 1950s and 1960s. It is still home to workers, and also open to visitors who are interested in the history of the town.

**Talbingo** is home to Tumut 3, the largest power station in the Snowy Mountains Scheme.

Cabramu

## HYDRO TASMANIA

Visitors can tour Trevallyn Power Station and Lake Margaret Power Station all year round, through tour operators. Other tours are held less regularly to other power stations.

# Glossary

**aneroid barometer** a surveying device that allows surveyors to accurately measure heights in rugged countryside

**antar** a large earth-carrying truck used on the Snowy Mountain Scheme

**bored** drilled or dug through the ground

**dam** a wall constructed to hold back a body of water

**Federal Government** the Government elected to govern Australia

**Hydro-electric power** electricity produced by the action of water turning a turbine attached to a generator

**loaders** machines that lift heavy loads of soil and rocks onto trucks

**ore** rock containing metal

**power station** a building that houses the apparatus or machinery to generate electricity

**renewable energy** energy from sources that replenish themselves

**slide rule** a scientific device often used by engineers to make accurate measurements

**smelting plant** a factory in which ores are heated to a liquid state in a blast furnace

**spillway** the main outlet of a dam, constructed to allow water to be released from behind the dam to the river or lake below

**tellurometer** an electronic surveying device that measures distances using sound waves

**zinc** a hard white metal

# Index